I0065152

SpaceX Fan COLORING BOOK

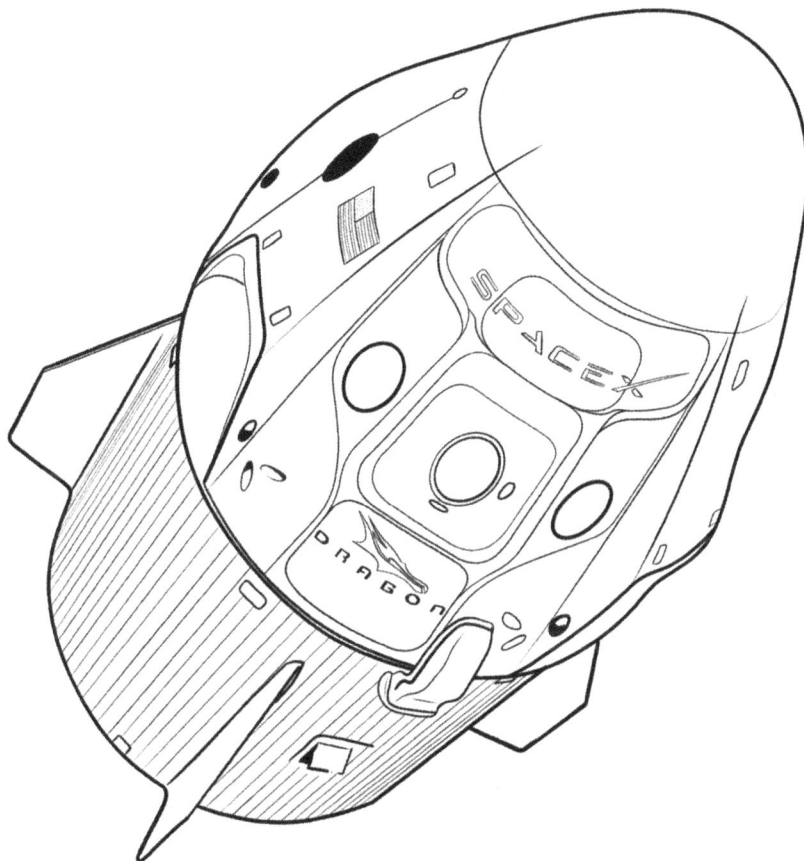

The Authoritative Space Coloring Book With SpaceX Starship, Dragon Capsule, Elon Musk, and More

Aero Maestro Books

SPACE-X was founded by Elon Musk in 2002 to revolutionize human space exploration by dramatically reducing the cost of delivering payloads (both people & cargo) to space.

In later interviews (in the 2018 timeframe), Elon said that at the founding he gave SpaceX a 10-20% chance of success as a company (meaning an 80-90% chance of failure). As you will learn in the following pages, they came very close. The early years saw one expensive and dramatic failure after another.

But eventually - with learning, persistence, and innovative engineering - SpaceX has become profoundly successful in its mission. Some of its "firsts" include:

- First private company to successfully launch, orbit, and recover a spacecraft (Dragon in 2010),
- First private company to send a spacecraft to the International Space Station (Dragon in 2012)
- First return-to-Earth landing of an orbital rocket (Falcon 9 in 2015)
- First re-use of an orbital rocket (Falcon 9 in 2017)
- First private company to deliver astronauts to the International Space Station (ISS) (Dragon 2 in 2020)

Enjoy immersing yourself in a new era of human space exploration thanks to SpaceX, Elon Musk, and all of the talented passionate people included in the SpaceX team - both within the company and in their external team of partners and stakeholders. Maybe it will include you some day! (As a fan of SpaceX with this book, you are taking an early small step!)

Fun fact: This book has been designed and produced by Aero Maestro, a company founded and led by an actual rocket scientist (an aerospace engineer).

If you enjoyed this book (or if you have any other constructive feedback to help us improve), we would be extremely grateful for your online review of this book. Of course, we hope you feel it's worth 5 stars but we appreciate your honest review most of all.

Elon Musk overlooking the remains of a Falcon 9R rocket after it exploded soon after liftoff on August 22, 2014.

Although this was a test flight, the first three production launches for SpaceX also ended in explosions or failures. Elon said later that they were only one more launch failure away from ending the company SpaceX.

Elon Musk at a press conference on June 28, 2015, following the explosion of a SpaceX Falcon 9 rocket intended to resupply the International Space Station. (It was also his 44th birthday.)

Give these SpaceX rockets some unique colors and designs!

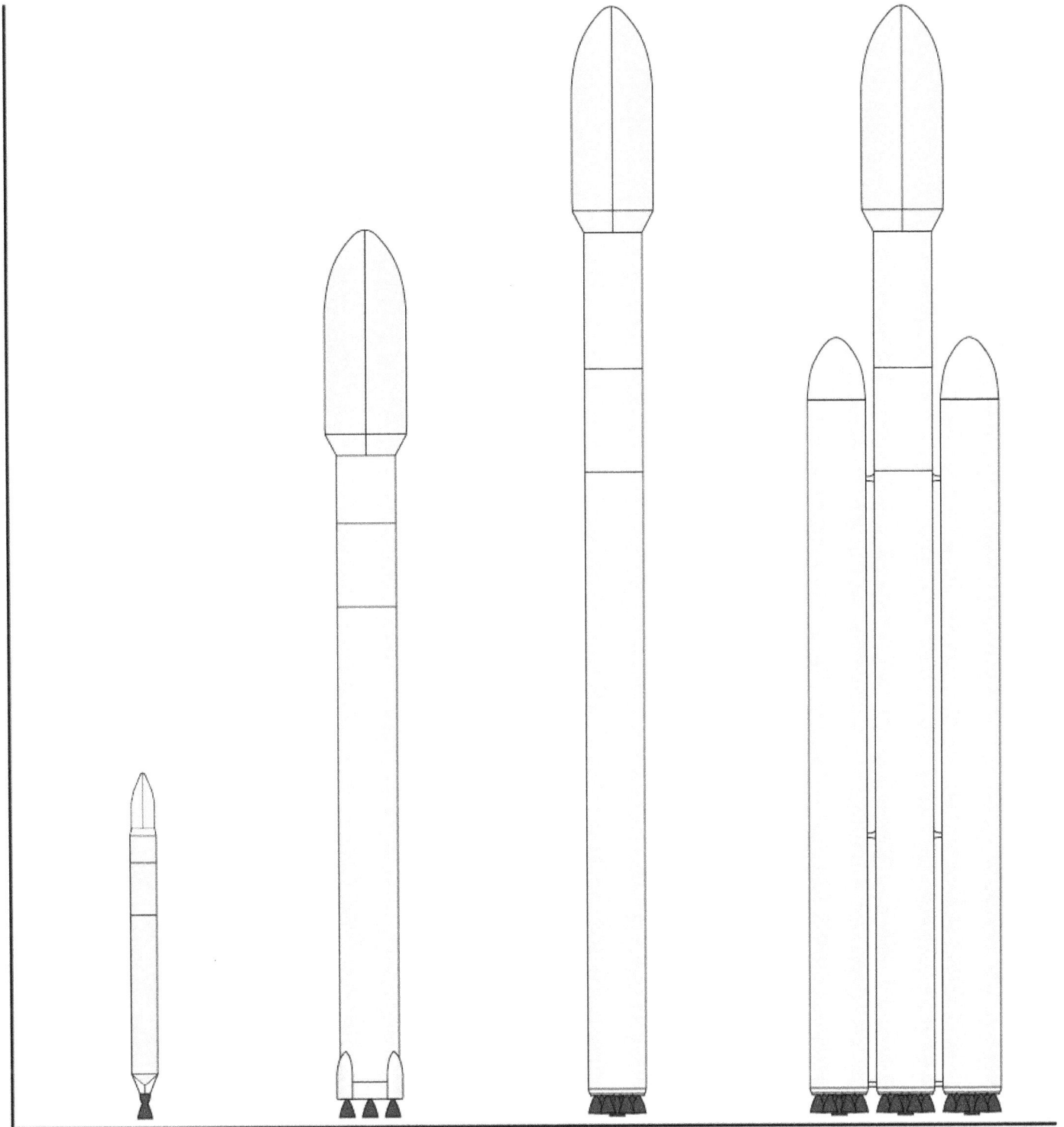

Falcon 1

Falcon 9
Block 1

Falcon 9

Falcon heavy

Elon watches with awe at the first launch of the Falcon 9 Heavy (F9R) rocket at Kennedy Space Center in Florida, USA, on February 6, 2018.

The historic landing on not one but TWO reusable rocket boosters from a SpaceX Falcon 9R rocket at Cape Canaveral, Florida, USA, in February 2018 (the same launch that contained Starman)

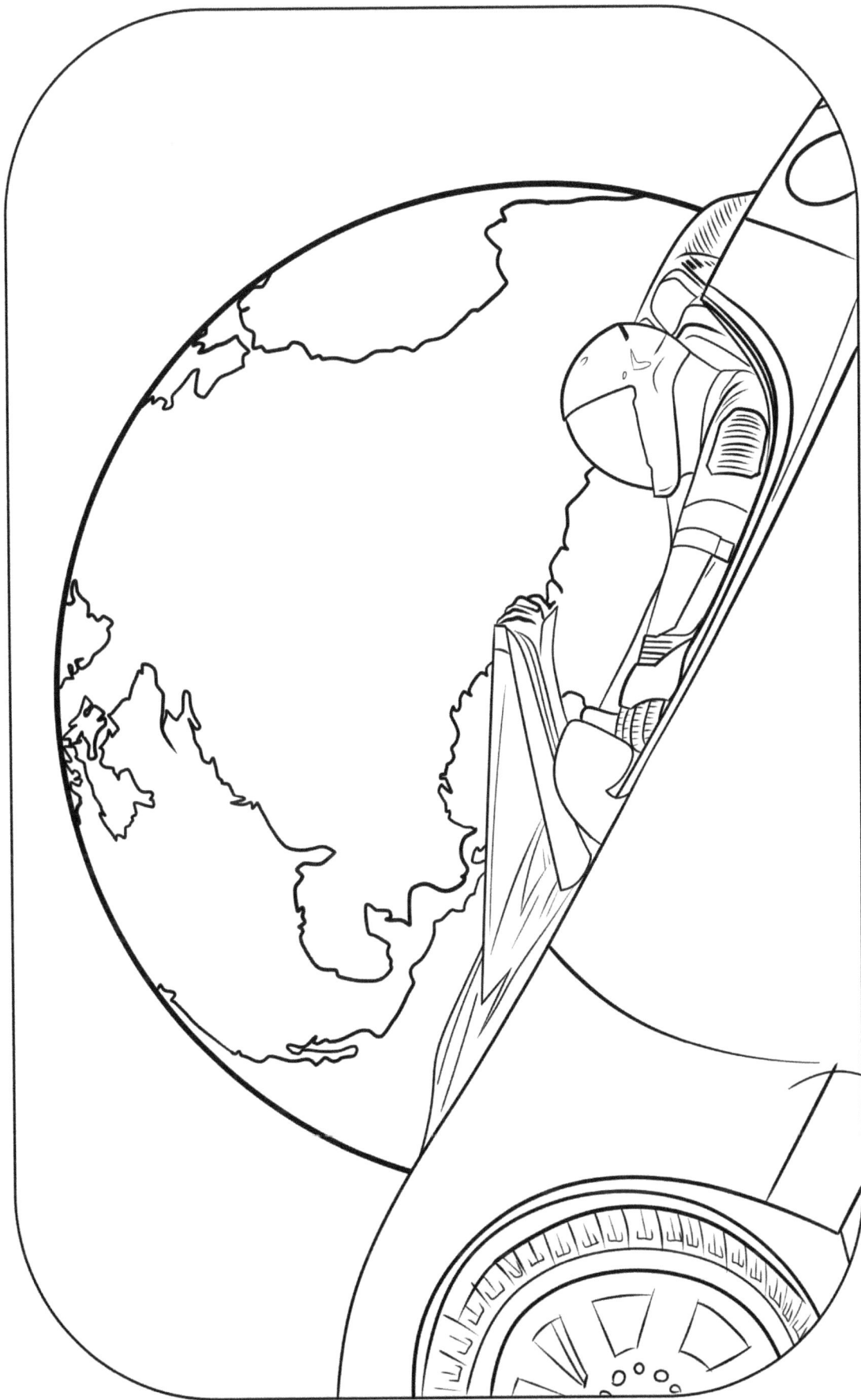

Starman onboard a Tesla Roadster. He is coasting towards Mars after being launched and released from a Falcon 9 rocket in February, 2018.

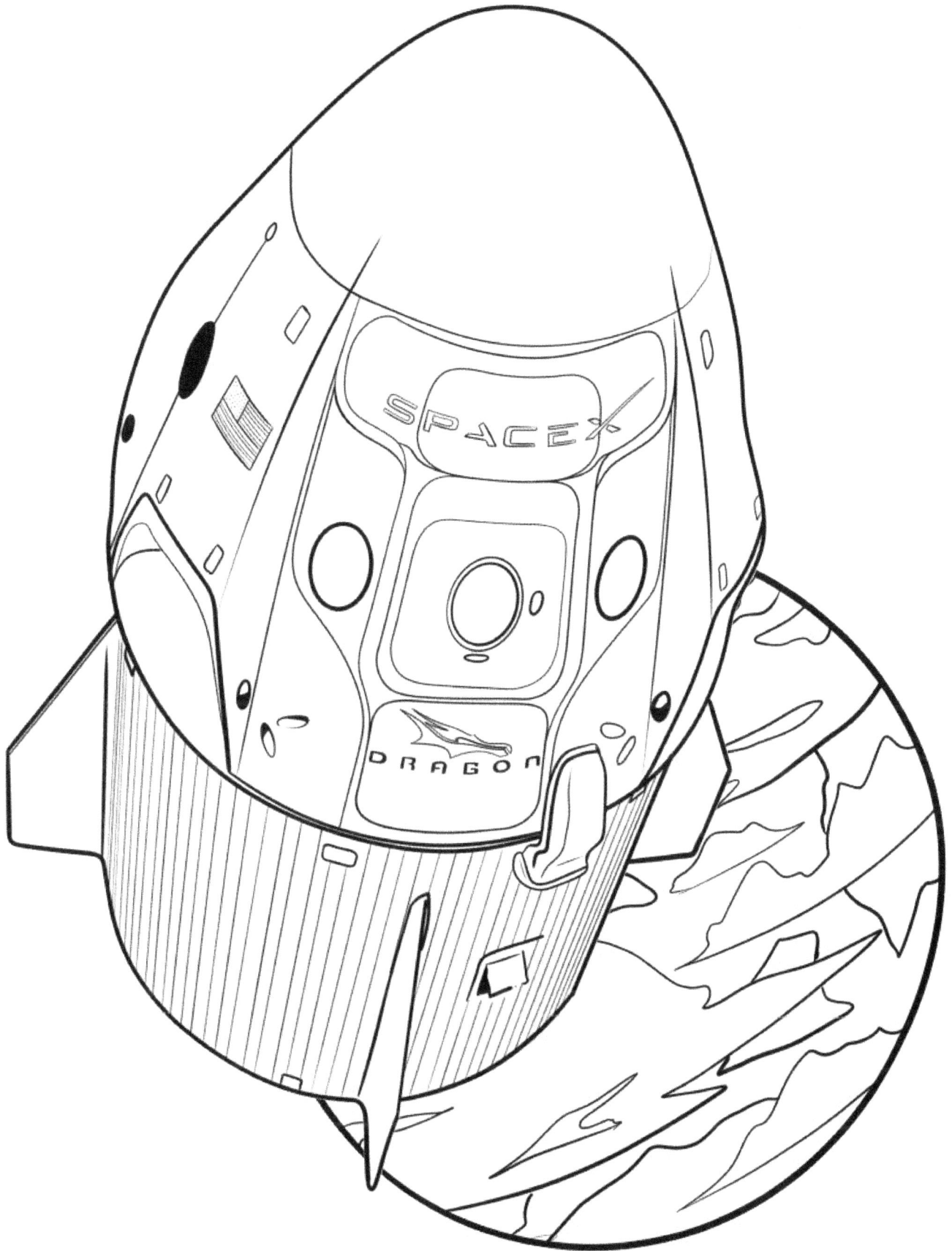

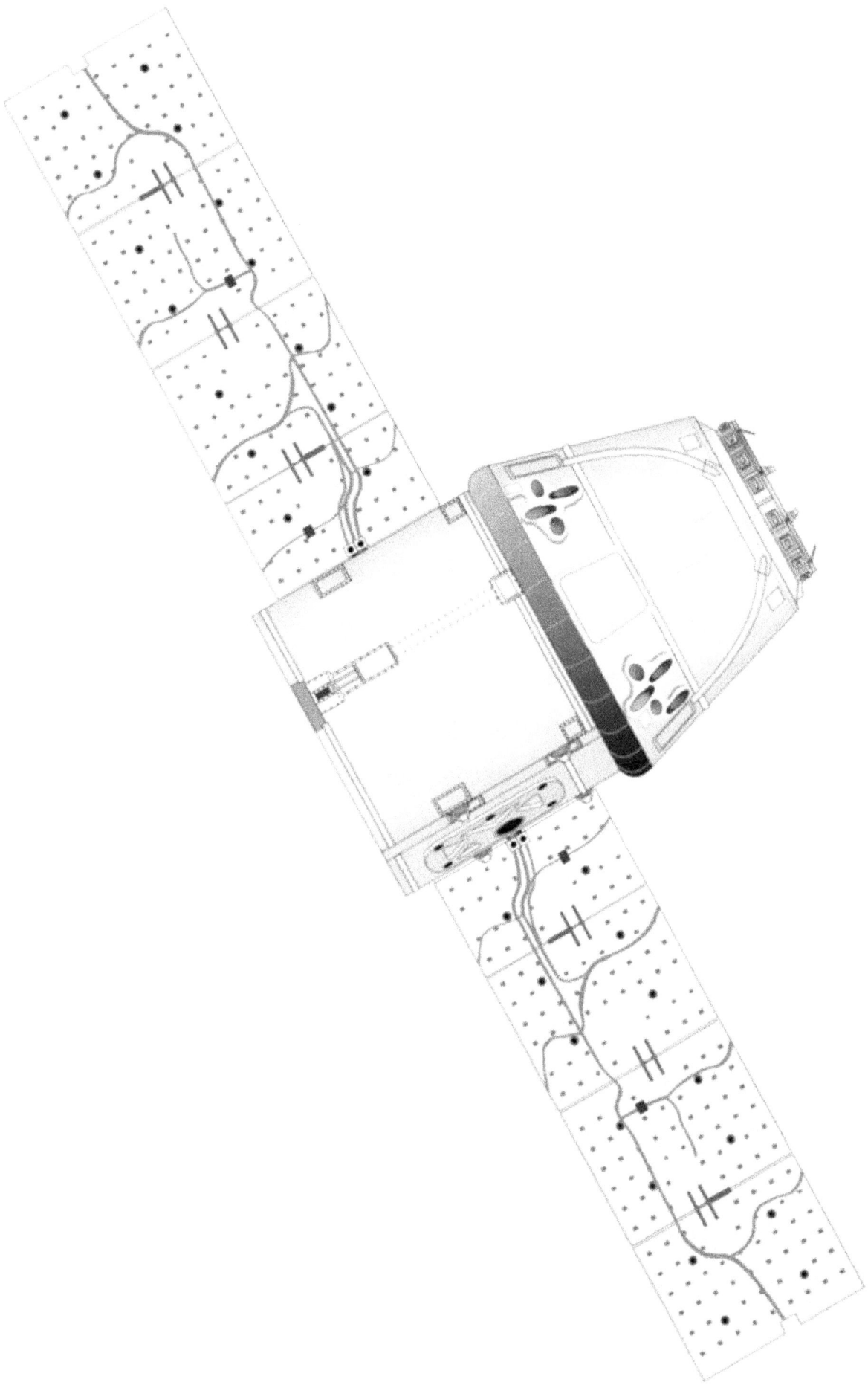

Elon Musk surveying the construction of the
first Gigafactory near Reno, Nevada, USA

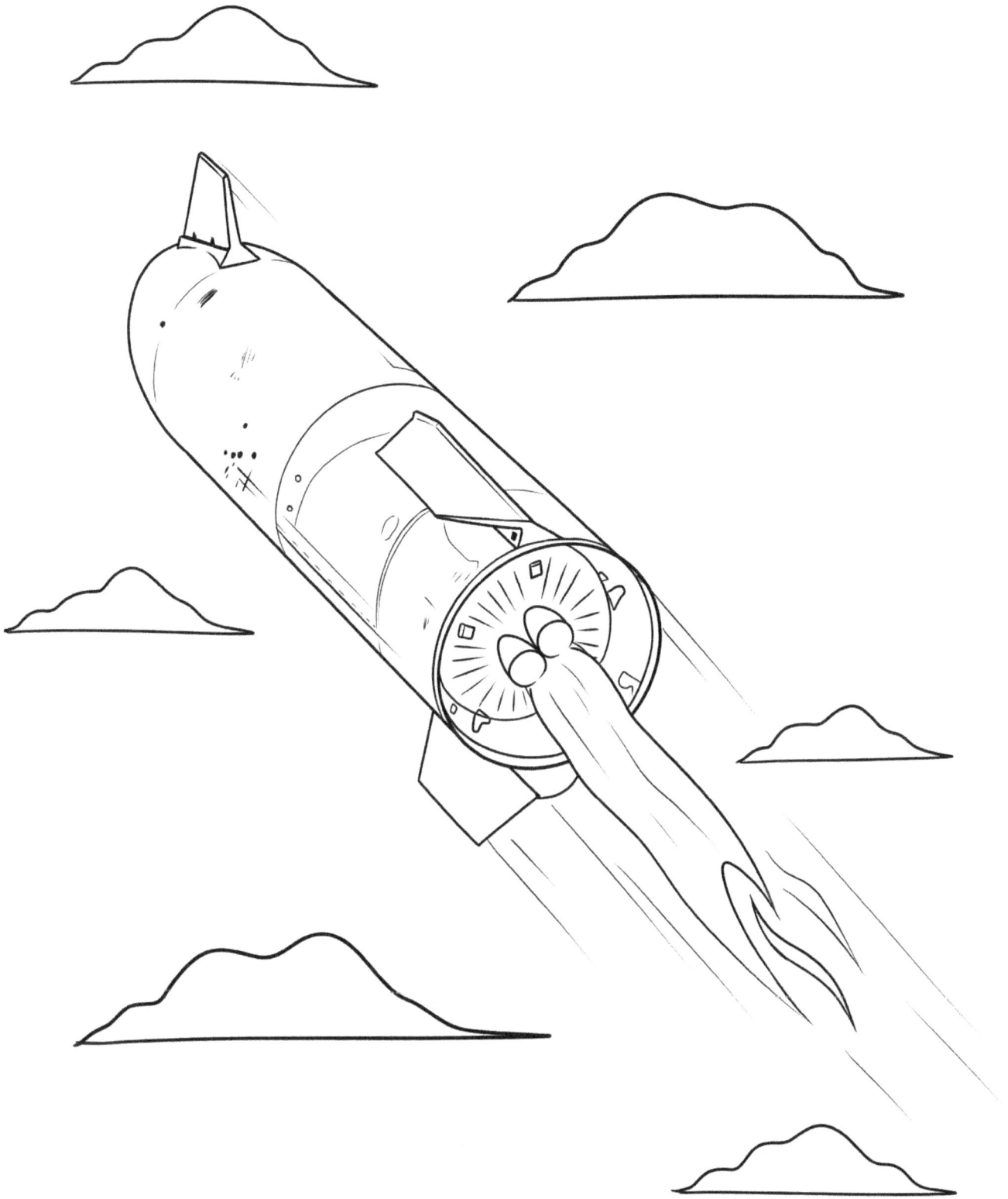

BONUS ADDITION:

SPACEX GAMES & PUZZLES!

(Keep reading to learn how you can get more like this.)

SPACEX WORD SCRAMBLE

1. C S X A E P   — — — — — —

2. R E O C T K   — — — — — —

3. N H L A U C   — — — — — —

4. L C N F O A   — — — — — —

5. K U L M S E N O   — — — —　　— — — —

6. I F R O L N A C I A   — — — — — — — — —

7. O D A I L F R   — — — — — — —

8. S E A E R B L U   — — — — — — — —

9. I I V N T N O O A N   — — — — — — — — — —

10. R P T A I E V   — — — — — — —

ELON ON PEOPLE

A	B	C	D	E	F	G	H	I	J	K	L	M	N	O	P	Q	R	S	T	U	V	W	X	Y	Z
																							1		

13   14 25 13 5 4   13 14   13 23   3 9 23 23 13 11 6 18   24 9 15

9 15 20 13 5 16 15 8   3 18 9 3 6 18   14 9   7 25 9 9 23 18

14 9   11 18   18 1(X) 14 15 16 9 15 20 13 5 16 15 8

Need another hint?
The word "PEOPLE" is one of the words in this cryptogram.

Cryptogram made thanks to:

www.KidZone.ws

Cryptogram

ELON ON PERSISTENCE

A	B	C	D	E	F	G	H	I	J	K	L	M	N	O	P	Q	R	S	T	U	V	W	X	Y	Z
				16																					

```
__  E  __  __  __  __  __  E  __  E     __  __     __  E  __  __
7   16 23 14 17 14 1  16 5  25 16    17 14    10 16 23 11
```

```
__  __  __  __  __  __  __  __  __       __  __  __     __  __  __  __  __  __     __  __  __
17 26 7  21 23 1  6  5  1         11 21 12    14 2  21 12 24 18    5  21 1
```

```
__  __  __  E     __  __     __  __  __  E  __  __     __  __  __     __  __  E
8  17 10 16    12 7     12 5  24 16 14 14    11 21 12    6  23 16
```

```
__  __  __  __  E  __     __  __     __  __  __  E     __  __
19 21 23 25 16 18    1  21     8  17 10 16    12 7
```

Need another hint?
You should find the topic of Elon's message in the cryptogram.

Cryptogram made thanks to:

www.KidZone.ws

Cryptogram

SUPER BONUS ADDITION:

SPaceX
TWO-PLAYER
Games!

(Keep reading to learn how you can get more like this.)

Time to play Dots and Boxes!

Who can make the most boxes on the Dragon Capsule solar panel?
Make more dots and boxes in the open space if you wish - or draw some other spacecraft, astronauts, or extraterrestrials flying nearby!

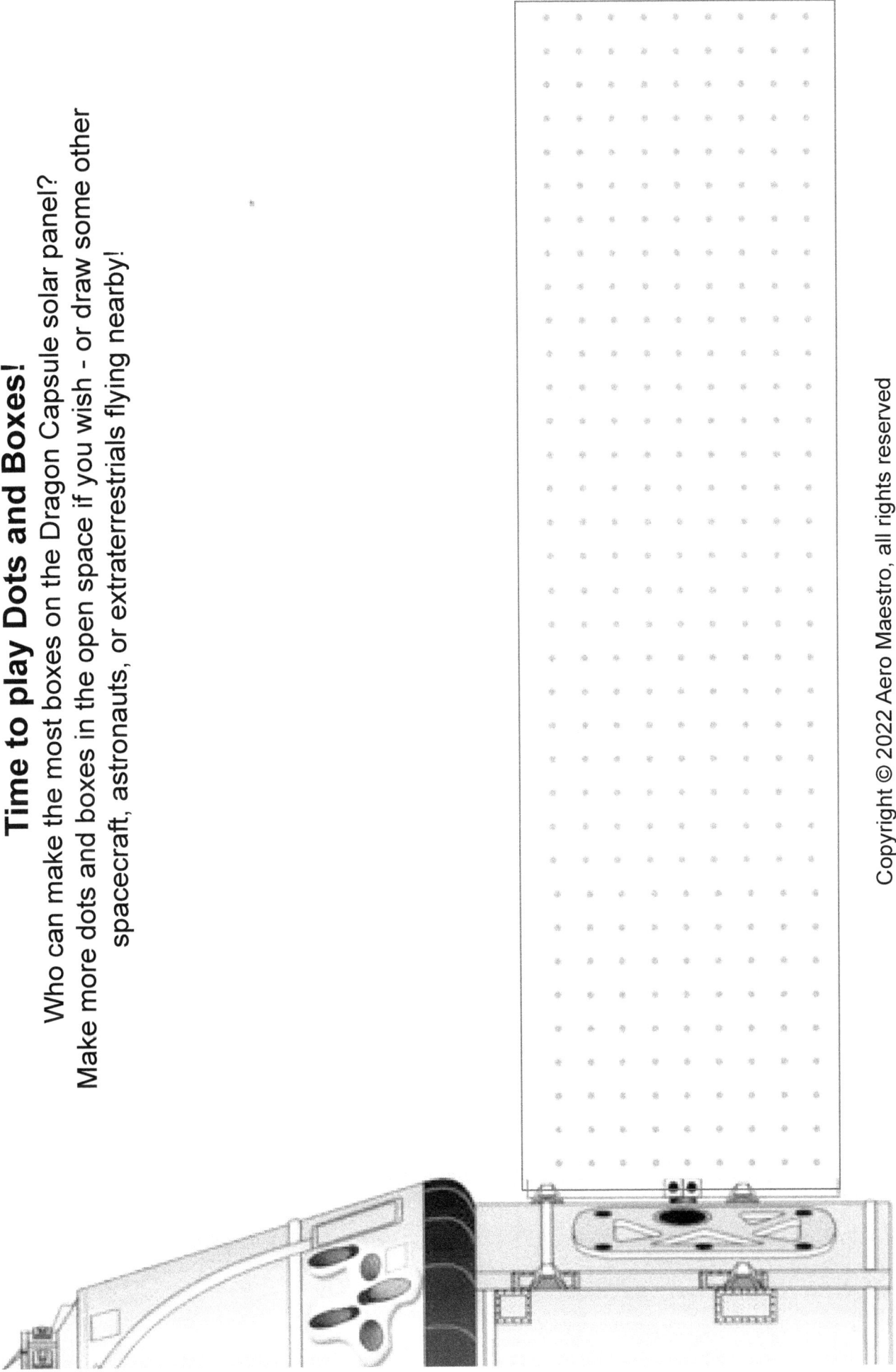

3-D Tic Tac Toe

Win with 4 in a row. Look out for those diagonals!

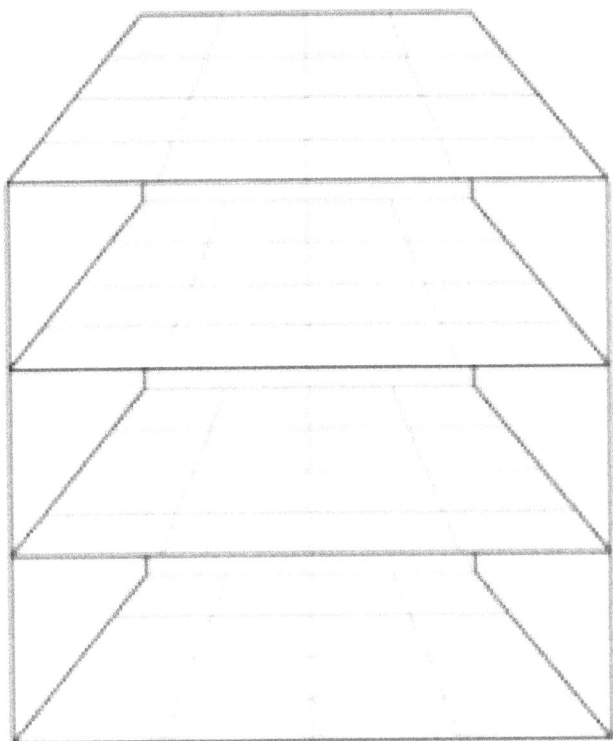

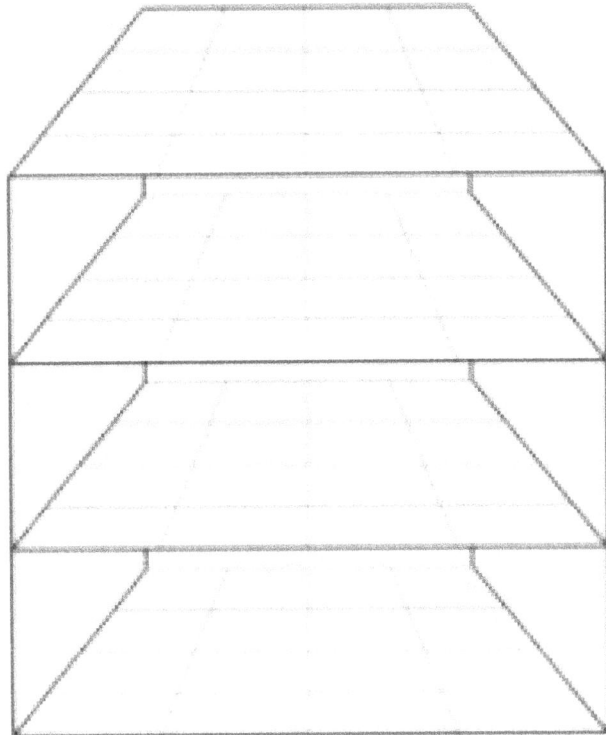

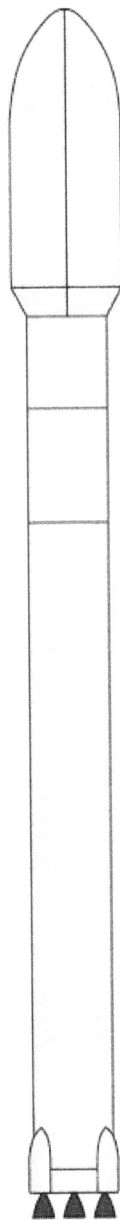

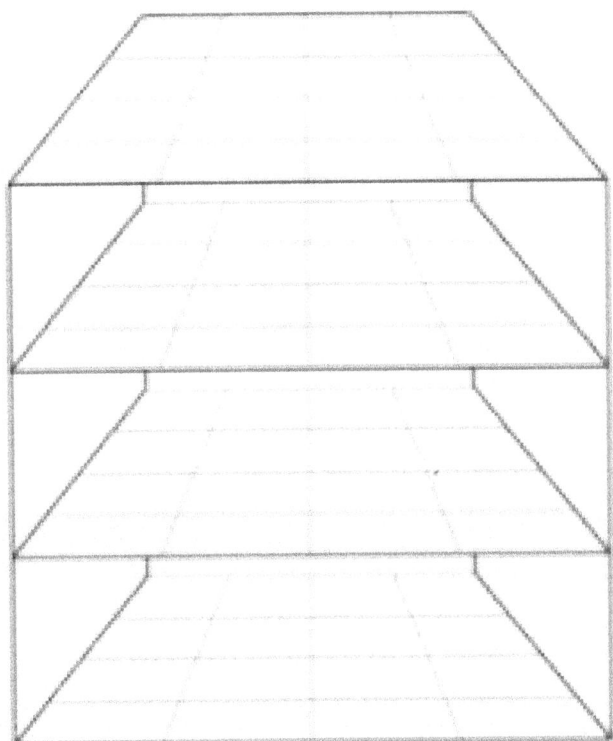

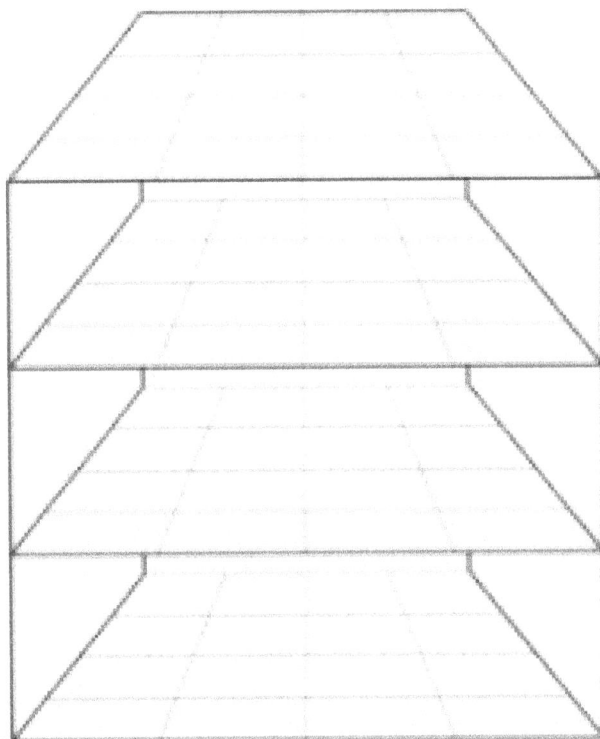

ANSWERS

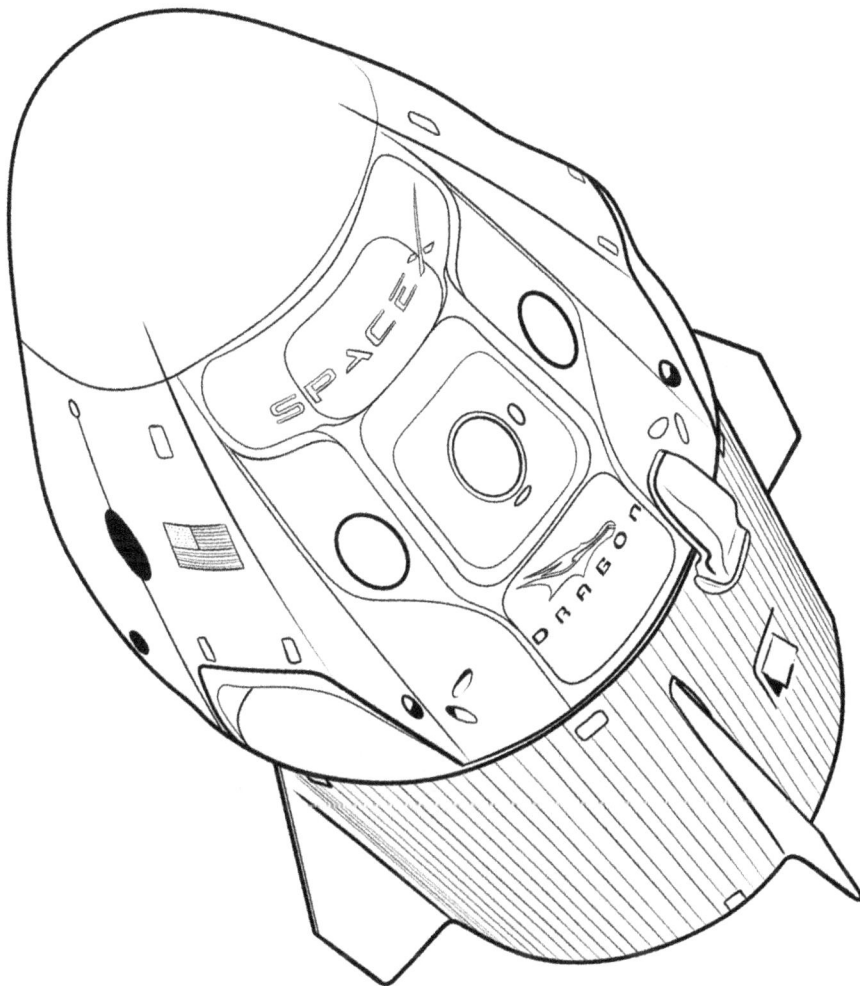

SPACEX WORD SCRAMBLE

1. C S X A E P   S P A C E X

2. R E O C T K   R O C K E T

3. N H L A U C   L A U N C H

4. L C N F O A   F A L C O N

5. K U L M S E N O   E L O N   M U S K

6. I F R O L N A C I A C A L I F O R N I A

7. O D A I L F R   F L O R I D A

8. S E A E R B L U   R E U S A B L E

9. I I V N T N O O A N I N N O V A T I O N

10. R P T A I E V   P R I V A T E

ELON ON PEOPLE

A	B	C	D	E	F	G	H	I	J	K	L	M	N	O	P	Q	R	S	T	U	V	W	X	Y	Z
16	11	7	20	18	24	17	25	13	10	4	6	21	5	9	3	22	15	23	14	12	2	26	1	8	19

```
  I       T  H  I  N  K        I  T       I  S        P  O  S  S  I  B  L  E        F  O  R
 13      14 25 13  5  4       13 14      13 23        3  9 23 23 13 11  6 18       24  9 15


  O  R  D  I  N  A  R  Y        P  E  O  P  L  E        T  O        C  H  O  O  S  E
  9 15 20 13  5 16 15  8        3 18  9  3  6 18       14  9        7 25  9  9 23 18


        T  O       B  E       E  X  T  R  A  O  R  D  I  N  A  R  Y .
       14  9      11 18      18  1 14 15 16  9 15 20 13  5 16 15  8
```

ELON ON PERSISTENCE

A	B	C	D	E	F	G	H	I	J	K	L	M	N	O	P	Q	R	S	T	U	V	W	X	Y	Z
6	4	25	18	16	19	8	2	17	13	15	24	26	5	21	7	22	23	14	1	12	10	3	20	11	9

```
P  E  R  S  I  S  T  E  N  C  E        I  S        V  E  R  Y
7 16 23 14 17 14  1 16  5 25 16       17 14       10 16 23 11


I  M  P  O  R  T  A  N  T  .        Y  O  U        S  H  O  U  L  D        N  O  T
17 26  7 21 23  1  6  5  1          11 21 12       14  2 21 12 24 18        5 21  1


G  I  V  E        U  P        U  N  L  E  S  S        Y  O  U        A  R  E
8 17 10 16       12  7       12  5 24 16 14 14       11 21 12        6 23 16


F  O  R  C  E  D        T  O        G  I  V  E        U  P  .
19 21 23 25 16 18        1 21        8 17 10 16       12  7
```

THANK YOU!

Thanks for purchasing this coloring book to immerse yourself in the world of SpaceX. We hope it brought you more enjoyment and appreciation for how fortunate we are to be alive today. SpaceX rockets are launching and delivering payloads to space and beyond on a regular basis. Then landing back on Earth so they can do it again soon!

Would you be willing to share your impressions of this book? We would love to get an online review from you (or the parent or teacher who gave you this book). We hope to hear from you there. Thanks very much!

WANT MORE?

We at **Aero Maestro** have created many other books that appeal to people like you. Look for Aero Maestro on Amazon or wherever you like to shop for books. You might find books like these:

- **Goodnight Moon Base**
- **Tesla Fan Puzzle & Activity Book**
- **Mad Words With Space Adventures: Silly Story Fill-ins**
- *And more on the way!*

Aero Maestro

We at **Aero Maestro** have more books planned and special offers for people who love advanced technologies, exciting innovations, and new worlds with human achievement. Would you love discounts on any of these books? The ability to request a coloring, puzzle, or activity book on a topic that you love?

EMAIL US HERE:
aero-maestro@protonmail.com

www.ingramcontent.com/pod-product-compliance
Lightning Source LLC
Chambersburg PA
CBHW081749200326
41597CB00024B/4446